The NSTA Reader's Guide to NEXT GENERATION SCIENCE STANDARDS

507.1073

The NSTA Reader's Guide to THE NEXT GENERATION SCIENCE STANDARDS

HAROLD PRATT

National Science Teachers Association

Claire Reinburg, Director
Jennifer Horak, Managing Editor
Andrew Cooke, Senior Editor
Amanda O'Brien, Associate Editor
Wendy Rubin, Associate Editor
Amy America, Book Acquisitions Coordinator

ART AND DESIGN
Will Thomas Jr., Director
Rashad Muhammad, Designer, Cover and Interior Design

PRINTING AND PRODUCTION
Catherine Lorrain, Director

NATIONAL SCIENCE TEACHERS ASSOCIATION
David L. Evans, Executive Director
David Beacom, Publisher

1840 Wilson Blvd., Arlington, VA 22201
www.nsta.org/store
For customer service inquiries, please call 800-277-5300.

Copyright © 2013 by the National Science Teachers Association.
All rights reserved. Printed in the United States of America.
16 15 14 13   4 3 2 1

NSTA is committed to publishing material that promotes the best in inquiry-based science education. However, conditions of actual use may vary, and the safety procedures and practices described in this book are intended to serve only as a guide. Additional precautionary measures may be required. NSTA and the authors do not warrant or represent that the procedures and practices in this book meet any safety code or standard of federal, state, or local regulations. NSTA and the authors disclaim any liability for personal injury or damage to property arising out of or relating to the use of this book, including any of the recommendations, instructions, or materials contained therein.

PERMISSIONS
Book purchasers may photocopy, print, or e-mail up to five copies of an NSTA book chapter for personal use only; this does not include display or promotional use. Elementary, middle, and high school teachers may reproduce forms, sample documents, and single NSTA book chapters needed for classroom or noncommercial, professional-development use only. E-book buyers may download files to multiple personal devices but are prohibited from posting the files to third-party servers or websites, or from passing files to non-buyers. For additional permission to photocopy or use material electronically from this NSTA Press book, please contact the Copyright Clearance Center (CCC) (*www.copyright.com*; 978-750-8400). Please access *www.nsta.org/permissions* for further information about NSTA's rights and permissions policies.

ISBN: 978-1-938946-06-6
e-ISBN: 978-1-938946-86-8

Cataloging-in-Publication data are available from the Library of Congress.

CONTENTS

About the Author vii

Preface, *by Karen L. Ostlund* ix

Introduction xi

Chapter 1: Getting Started 1

Chapter 2: The *Framework* and History 7

Chapter 3: Introduction to the *NGSS* 11

Chapter 4: Moving From the *NGSS* to Instruction 17

Chapter 5: A Guide for Leading a Study Group on the Total Curriculum 21

Chapter 6: Implementation 27

References 35

Index 37

ABOUT THE AUTHOR

Harold Pratt, a career district science coordinator and a former NSTA president, is a consultant working in all areas of science education. He was a senior program officer at the National Research Council during the development of the *National Science Education Standards.* He is the recipient of NSTA's Robert H. Carlton Award and the coauthor of a number of science textbooks and articles.

PREFACE

NSTA fully supports the *Next Generation Science Standards (NGSS)* and believes the *NGSS* offers an extraordinary opportunity to move science education forward in the 21st century. NSTA is committed to supporting the science education community and offers many tools that promote better understanding of the *NGSS*, including *The NSTA Reader's Guide for the Next Generation Science Standards,* an essential companion to the *Next Generation Science Standards* as states, districts, and schools prepare to adopt and implement the new standards. One of NSTA's strategic goals is to advocate for the central role of science education to benefit students and society. To that end, NSTA provides high-quality products and services—print and electronic publications, e-newsletters, online and face-to-face professional development (including institutes and web seminars), conferences, and symposia that focus on the *NGSS*—to guide science educators along the continuum from awareness to adoption to implementation. Lean on our library of resources as you develop instructional and assessment strategies that address the *NGSS*. NSTA continues to develop exemplary training manuals and materials to support professional development efforts. This *Reader's Guide* is a requisite piece in these efforts.

—**Karen L. Ostlund**
2012–2013 NSTA President

INTRODUCTION

The *Next Generation Science Standards (NGSS)* comprises many parts with many purposes, as evidenced by the contents of the *NGSS* website listed in Figure I.1. This *Reader's Guide* is designed to help you navigate and understand this array of parts, interpret the standards, and take the first steps toward putting the standards into practice.

FIGURE I.1. *NGSS* table of contents of the total *NGSS* document

NGSS Front Matter
NGSS Structure
Appendices to the *NGSS*
    A. Conceptual Shifts
    B. Responses to May Public Feedback
    C. College and Career Readiness (coming soon)
    D. All Standards, All Students (coming soon)
    E. Disciplinary Core Idea Progressions
    F. Science and Engineering Practices
    G. Crosscutting Concepts
    H. Nature of Science (coming soon)
    I. Engineering Design in the *NGSS*
    J. Science, Technology, Society, and the Environment
    K. Model Course Mapping in Middle and High School (coming soon)
    L. Connections to CCSS-Mathematics (coming soon)
    M. Connections to CCSS-ELA Literacy (coming soon)
Commonly Used Abbreviations
Why Standards Matter
Public Attitudes Toward Science Standards

Although there are many ways to navigate and make sense of the *NGSS*, this guide suggests a process that should be most helpful to you. By following the sequence of chapters in the guide, you will come to understand the various pieces listed in Figure I.1 in a logical and cohesive manner. The *Reader's Guide* is designed to help you move along the path in Figure I.2 from the beginning stages of awareness and understanding the nature of the *NGSS*, to the next stage of exploring how the *NGSS* translates to instruction, and finally to the stage of early planning for implementation.

FIGURE I.2. The path of progress in learning to use the *NGSS*

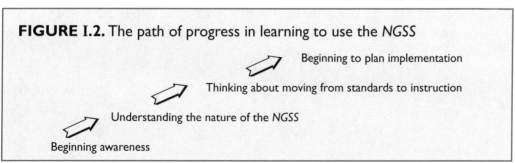

By the time you finish all six chapters, you will be on your way to a deep understanding of the *NGSS* and the promise it holds for science education. It's an important journey to navigate, holding the potential for your own professional growth and for a better education for your students. Future NSTA publications will pick up where this guide leaves off with many more details about using and implementing the *NGSS*.

Not everyone interested in studying the *NGSS* will want or need, at least immediately, to proceed the entire length of the path in Figure I.2. Your depth of learning and engagement will depend on your interest or role. If you are responsible for coordination and supervision at the state, regional, or district level, this guide may be a useful starting point in your understanding and planning before moving on to more comprehensive resources.

If you are a classroom teacher, the first four chapters may be the focus of your early reading. Chapter 4 will help you better understand how the *NGSS* can be used to plan instruction or instructional materials. This guide is not a comprehensive how-to manual, but rather a framework to provide what we might call an operational understanding of the *NGSS*. Rather than just read the standards, use this guide to stimulate in-depth thinking about how to use the *NGSS* for instructional purposes. In doing so, you will develop a better understanding of the components of the standards and how they can be used.

Although it sometimes occurs, teachers should never be handed standards, including the *NGSS*, and then be expected to translate them into classroom practice on their own. Teachers are certainly capable of doing so, but the time and effort required of such a task is beyond the scope of the normal teaching assignment and the time and resources allocated for the job. This guide will help the entire science education team in a district or state understand the extent of the tasks—outlined in chapters 4, 5, and 6—that need to be accomplished. The actual work involved calls for a significant team effort and adequate resources.

The information and messages in this *Reader's Guide* are in many ways the product of an NSTA team that has worked together for three years to review the *Framework* and then several drafts of the *NGSS*. Cindy Workosky coordinated the process of reviewing and providing feedback to the National Research Council (NRC) during the development of *A Framework for K–12 Science Education* (NRC 2012) and to Achieve Inc. during the review of the five *NGSS* drafts. Shortly after the *Framework* was released, Ted Willard joined the NSTA staff, leaving the American Association for the Advancement of Science (AAAS) after twelve years of work on the *Benchmarks for Science Literacy* and *Atlas of Science Literacy*, to help lead the review and feedback process of the *NGSS* drafts. More recently, as the work progressed to engaging NSTA membership in professional development, Zipporah Miller has lent her time and experience to this effort.

I have had the privilege of working with this team from the beginning and gaining from their knowledge, expertise, and experience as we considered the many issues and processes shared in this *Reader's Guide*. I am grateful for what I have learned from their collegial sharing and insights. This is their work as well as mine.

—Harold Pratt

CHAPTER 1
GETTING STARTED

Welcome to the *Next Generation Science Standards (NGSS)*!

The new standards promote a vision that the instruction and the learning your students experience should change significantly. An early piece of front matter in the *NGSS* that lists the conceptual shifts called for in the *NGSS* provides a brief overview of the changes:

- K–12 science education should reflect the real-world interconnections in science.
- The *Next Generation Science Standards* are student outcomes and are explicitly NOT curriculum.
- Science concepts build coherently across K–12.
- The *NGSS* focuses on deeper understanding and application of content.
- Science and engineering are integrated in science education from K–12.
- Science standards coordinate with *Common Core State Standards* in English language arts and mathematics.

You may be reading this guide shortly after the new standards are released or picking it up many months later with the intention of getting better acquainted with the *NGSS* and the changes that should occur. In any case, this guide is designed to assist you and your team in gaining a deep understanding of the new standards and will supplement the many front matter documents and appendices provided with the *NGSS* itself. These documents should be carefully read as you progress through this guide and its many suggestions for action.

The *Reader's Guide* contains a series of suggested recommendations for action, starting with a plan to intensely study the *NGSS*—first by becoming familiar with the architecture of the document, then by thinking about the development of instructional strategies and the corresponding materials as a way of gaining a deeper and more "operational" understanding of the standards and how they should be used. These suggestions give you latitude to adjust them for your needs and rate of progress. This guide has been created as a part of an array of resources from NSTA to support you and your colleagues. These resources are listed at the end of this chapter.

Getting Started

A Checklist

The following checklist of activities will help you and your team carry out the process of first gaining an understanding of the *NGSS* and then developing a plan for putting the standards into practice. The checklist is followed by a few details and suggestions for each item with references to chapters in this *Reader's Guide*.

- Determine your state's or district's plan for implementation and timeline.
- Form a team or study group.
- Collect resource materials from NSTA and other sources.

- Follow a plan to study the *NGSS* using the suggestions in this guide.
- Create a plan for putting the *NGSS* into practice.
- Use the *NGSS* in limited trial situations.

More Details

Determine Your State or District's Plan for Adoption and Implementation Decisions and Timeline

Although the release of the *NGSS* in the spring of 2013 may appear to be the starting point for planning the adoption and implementation of the new standards in your state or school district, in many cases, an adoption decision may not yet have been made or work toward adoption might already have begun. In any event, your first important step is to determine if, how, and when adoption decisions are made in your school, district, or state. (If this is done at the state level, make your inquiries to the appropriate person such as the science coordinator or commissioner of education.) Even if an adoption date is known, it will most likely be several years before significant implementation is required in your school, district, or state. So, determining these dates is a critical first step for you.

If your state or district does not adopt the *NGSS* or at least has no plans to do so, you should still encourage your school to use the best practices of the *NGSS*. The content in the *Framework* and the *NGSS* represents current research and best practices and thinking in science education and will provide significant insights about improving science teaching and learning.

Form a Team or Study Group

In almost all circumstances, planning and implementing the standards will require a team effort. (Please also note the importance of a team effort discussed in the Introduction.) To share the myriad tasks suggested in this *Reader's Guide,* you should involve teachers and educators at all levels and in all science content areas. See Chapter 5 in this guide for a discussion of how many team members provide the maximum coverage for dealing with the entire *NGSS* document.

Collect Resources From NSTA and Other Sources

See the list of NSTA resources at the end of this chapter. Collecting resources is easy and can start early in your learning process. As additional resources become available, continue to collect what you need while also working through the other steps in your process.

Follow a Plan to Study the *NGSS*

Based on suggestions later in this section, develop a tentative plan for an intensive study of the *NGSS* with dates, responsibilities, and a budget. Seek and attain approval as required. However, don't overplan at this point. Get started and continue to plan as you proceed. Unless you are required to submit a complete long-range plan, a dynamic planning process for such a new effort is advisable. Plan, get started, review and evaluate progress, revise the plan, and continue to work. This cycle will allow you to get started with early results as you continue the planning process.

The place to start in understanding and using the *NGSS* is the *Framework for K–12 Science Education.* The *Framework* provides a vision for K–12 science education and detailed information for the science and engineering practices, disciplinary core ideas, and crosscutting concepts. The *Framework* developed by the NRC was the first in a two-step process followed by the development of the *NGSS* by Achieve Inc. using the specifications outlined in the *Framework*. Chapter 2, The *Framework* and History, will be a useful resource in this process. Chapter 3, Introduction to the *NGSS*, provides a very detailed process for understanding and interpreting the complex structure or architecture of the standards.

There is no better way to understand and begin applying the standards than to use them in a limited way to design or plan an instructional sequence and the related instructional materials. Chapter 4, Moving From the *NGSS* to Instruction, is designed to take you through this exploratory process. By doing so, your team will gain a better understanding of how the standards can be translated into instruction. The evaluation of extant instructional materials and considering how to adapt them will also be included.

While the procedure outlined in Chapter 4 offers insight into how one standard can be translated into instruction, the process outlined in Chapter 5, A Guide for Leading a Study Group on the Total Curriculum, provides a procedure for a team to gain in-depth understanding of the entire *NGSS*. It is a modification of a process documented elsewhere (NSTA 2013a) during the review of the *NGSS* public drafts.

Create a Plan for Putting the *NGSS* Into Practice

Planning for implementation should begin as soon as decisions have been made about the adoption of the *NGSS*. Think of the *NGSS* as representing the starting and concluding point of your planning. It outlines the learning goals for your instructional program and a description of how to evaluate student success. Your team's task is to plan and implement the strategies of instruction, instructional materials, assessments, curriculum, and professional development that must take place between these two points. Chapter 6, Implementation, briefly outlines the components of a plan but is not a detailed set of instructions. It is designed to get your team thinking and started. Note: Achieve, Inc. has published a document (Achieve 2013) for state leaders who plan to adopt and implement the *NGSS*; this document could be helpful for your team.

Use the *NGSS* in Limited Trial Situations

If the adoption and implementation timeline allows, the impact on classrooms can be phased in over several years. If at all possible, begin with a trial use in a few classrooms with only a small portion of the standards being used.

This Guide Is Only the Beginning

This *Reader's Guide* is designed to get you started using the *NGSS* to improve the quality of science education in your school, district, or state. Other NSTA resources are available, or soon will be, to support you and your team in going further with the implementation of the *NGSS*.

Let's get started.

NGSS@NSTA Resources

Since 2011, NSTA has been helping teachers prepare for the *Next Generation Science Standards* by helping them better understand the foundation on which it is built—the *Framework for K–12 Science Education*. The result is a robust collection of print, online, and in-person resources.

With the release of the *NGSS*, NSTA is developing professional development focused on helping teachers learn more about the new standards, understand more deeply the content and architecture of the new standards, and translate the standards into classroom practice.

An essential starting place is NGSS@NSTA, located at *www.nsta.org/ngss*. From web seminars to books and journal articles, from short courses to institutes, our resource page always has something new to help teachers realize the vision of the *NGSS*. The following is just a sampling of what you will find there.

Books

Translating the NGSS *for Classroom Instruction* (summer 2013)
This guide by Rodger Bybee, an *NGSS* writer and colead for life sciences, answers questions about translating standards into classroom practices and gives insights about reforming curriculum for schools, districts, and states.

Science for the Next Generation: Preparing for the New Standards
This multicontributor book of essays and mini lessons, developed in conjunction with the Science Teachers Association of New York State, is aimed at elementary school teachers preparing for the *NGSS*.

The NSTA Reader's Guide to A Framework for K–12 Science Education: Practices, Crosscutting Concepts, and Core Ideas
This handy *Reader's Guide* unpacks the three key dimensions of the *Framework* allowing teachers, administrators, curriculum developers, university professors, and others to more easily grasp how the *NGSS* differs from previous standards. For each chapter of the *Framework*, the guide offers an overview with a brief synopsis of key ideas, an analysis of what is similar to and what is different from the *NSES*, and suggested actions to help readers understand and prepare for the *NGSS*.

Journal Series: Exploring the Science *Framework* and Preparing for the *NGSS*

Learn from the experts, including many writers of the *NGSS* and the *Framework*, in an impressive collection of journal articles that explore the *Framework* and its key dimensions.

Science and Engineering Practices

- "Scientific and Engineering Practices in K–12 Classrooms," by Rodger W. Bybee
- "Exploring the Science Framework: Engaging Learners in Scientific Practices Related to Obtaining, Evaluating, and Communicating Information," by Philip Bell, Leah Bricker, Carrie Tzou, Tiffany Lee, and Katie Van Horne
- "Engaging Students in the Scientific Practices of Explanation and Argumentation," by Brian J. Reiser, Leema K. Berland, and Lisa Kenyon
- "Engaging Students in Scientific Practices: What Does Constructing and Revising Models Look Like in the Science Classroom?" by Joseph Krajcik and Joi Merritt

Disciplinary Core Ideas
- "The *Next Generation Science Standards* and the Life Sciences: The Important Features of Life Science Standards for Elementary, Middle, and High School Levels," by Rodger W. Bybee
- "The *Next Generation Science Standards:* A Focus on Physical Science," by Joseph Krajcik
- "The *Next Generation Science Standards* and the Earth and Space Sciences: The Important Features of Earth and Space Science Standards for Elementary, Middle, and High School Levels," by Michael E. Wysession

Crosscutting Concepts
- "The Second Dimension—Crosscutting Concepts," by Richard A. Duschl

Engineering
- "Core Ideas of Engineering and Technology," by Cary Sneider

Connections to Common Core
- "Exploring the Science Framework: Making Connections in Math With the *Common Core State Standards*," by Robert Mayes and Thomas R. Koballa Jr.

Web Seminars Focused on the NRC *Framework*

NSTA hosted two series of web seminars for science educators—one designed to help them better understand the practices described in the *Framework* and one intended to explore the crosscutting concepts. The archived webinars are available for viewing at *www.nsta.org/ngss*.

Science and Engineering Practices

These web seminars focused on helping teachers understand the key elements of the practices, how the practices work together, and what the use of the practice looks like in the classroom.

- "Obtaining, Evaluating and Communicating Information," by Philip Bell, Leah Bricker, and Katie Van Horne
- "Engaging in Argument From Evidence," by Joseph Krajcik
- "Constructing Explanations and Designing Solutions," by Katherine McNeill and Leema Berland
- "Using Mathematics and Computational Thinking," by Robert Mayes and Bryan Shader
- "Analyzing and Interpreting Data," by Ann Rivet
- "Planning and Carrying Out Investigations," by Rick Duschl
- "Developing and Using Models," by Christina Schwarz and Cynthia Passmore
- "Asking Questions and Defining Problems," by Brian Reiser
- "Using the *NGSS* Practices in the Elementary Grades," by Heidi Schweingruber and Deborah Smith
- "Engineering Practices in the *Next Generation Science Standards*," by Mariel Milano
- "Connections Between Practices in the *NGSS*, Common Core Math, and Common Core ELA," by Sarah Michaels

Crosscutting Concepts

This web seminar series explores the crosscutting concepts to provide K–12 teachers with strategies for implementing the *NGSS* in the classroom.

- "Patterns," by Kristin Gunckel
- "Cause and Effect: Mechanism and Explanation," by Tina Grotzer
- "Scale, Proportion, and Quantity," by Amy Taylor and Kelly Riedinger
- "Energy and Matter—Flows, Cycles, and Conservation," by Andy Anderson

CHAPTER 2
THE *FRAMEWORK* AND HISTORY

The *Framework* as Prerequisite for Studying the *Next Generation Science Standards*

If you haven't done so already, study—at a minimum—the portions of the *Framework* discussed below. The best preparation for the *NGSS* is a thorough understanding of the *Framework*. A brief overview of those sections, as well as the history of the *Framework*'s development follows.

Foreword. You will find a brief description of the history of the *Framework*. Please note that though it is never stated, the funding for the *Framework* and the subsequent work on the *NGSS* was provided by the Carnegie Corporation. No federal funds were used in the process.

Summary. This is a brief overview of the *Framework*.

<u>Part I: A Vision for K–12 Science Education</u>

Chapter 1: A New Conceptual Framework. This chapter outlines the vision of the *Framework* and, ultimately, the *NGSS*.

Chapter 2: Guiding Assumptions and Organization of the Framework. As the title implies, this chapter provides the assumptions about learning on which the *Framework* and *NGSS* are based.

<u>Part II: Dimensions of the Framework</u>

Chapter 3: Dimension 1: Science and Engineering Practices. This chapter offers the rationale for the shift from "inquiry" to "practices," an extensive discussion of the eight practices, and a description of the progression that students should be able to move through in each practice from the early grades through grade 12. Appendix F in the *NGSS* document also discusses the purpose and nature of the practices.

Chapter 4: Dimension 2: Crosscutting Concepts. The nature and purpose of the crosscutting concepts is outlined with an extensive discussion of the seven concepts. The progression of understanding of the concepts by students from the early grades through grade 12 is described. Appendix G in the *NGSS* document also discusses the purpose and nature of the crosscutting concepts.

Chapter 5: Dimension 3: Disciplinary Core Ideas—Physical Sciences

Chapter 6: Dimension 3: Disciplinary Core Ideas—Life Sciences

Chapter 7: Dimension 3: Disciplinary Core Ideas—Earth and Space Sciences

Chapter 8: Dimension 3: Disciplinary Core Ideas—Engineering, Technology, and Applications of Science

These four *Framework* chapters outline the core ideas of science content with endpoints of understanding described for grades 2, 5, 8, and 12. The progression of these core ideas is also discussed in Appendix E of the *NGSS* document.

Part III: Realizing the Vision

Chapter 9: Integrating the Three Dimensions. This chapter addresses the need for integration of the practices, crosscutting concepts, and core ideas—the most innovative characteristic of the *NGSS*. **Reading this chapter in detail is critical before proceeding to a study of the *NGSS*.**

Note: The *Framework* in a paperback format can be purchased from NSTA at *www.nsta.org/store*. It can also be read or downloaded in PDF format at *www.nap.edu*.

Developing the NGSS: A Brief History

The genesis of the *NGSS* can be traced back to 2007, when the Carnegie Corporation of New York and the Institute for Advanced Study Commission on Mathematics and Science Education concluded in a report (Carnegie 2007) that "the nation's capacity to innovate for economic growth and the ability of American workers to thrive in the global economy depend on a broad foundation of math and science learning, as do our hopes for preserving a vibrant democracy." This report and the subsequent funding from the Carnegie Corporation to the National Research Council (NRC) and later to Achieve Inc. resulted in the development of *A Framework for K–12 Science Education* and then the *Next Generation Science Standards*.

The following paragraph, which describes the history and the partners who have contributed to the development of the new standards, was adapted from text in the front matter of the *NGSS* (Achieve 2013):

> The National Academy of Sciences, Achieve, the American Association for the Advancement of Science, and the National Science Teachers Association have been partners on a two-step process to develop the *Next Generation Science Standards*. The first step of the process was led by the National Academies of Science. On July 19, 2011, the National Research Council, the functional staffing arm of the National Academy of Sciences, released *A Framework for K–12 Science Education*. The *Framework* was a critical first step because it is grounded in the most current research on science and science learning, and it identifies the science all K–12 students should know. The second step in the process was Achieve Inc.'s development of the new standards based on the content in the NRC *Framework*. The standards have undergone numerous reviews by the partners and 26 lead states as well as during two public comment periods before their release in April 2013.

The Partners

National Academies—which comprises the National Academy of Sciences, the National Academy of Engineering, and the Institute of Medicine—is a nongovernmental organization commissioned to advise the nation on scientific and engineering issues. It is composed of distinguished scholars engaged in scientific and engineering research dedicated to furtherance of science and technology and to the use of that science and technology for the general welfare. The NRC is the principle operating agency of the National Academies.

The American Association for the Advancement of Science (AAAS) is an international nonprofit organization dedicated to advancing science around the world by serving as an educator, leader, spokesperson, and professional association. In addition to organizing membership activities, AAAS publishes the journal *Science*—as well as many scientific newsletters, books, and reports—and spearheads programs that raise the bar of understanding for science worldwide. AAAS is the home of Project 2061, the developers of *Science for All Americans*, *Benchmarks for Science Literacy,* and *Atlas of Science Literacy*.

Achieve Inc. is an independent, bipartisan, nonprofit education reform organization. At the 1996 National Education Summit, a bipartisan group of governors and corporate leaders created an organization dedicated to supporting standards-based education reform efforts across the states. Achieve is the only education reform organization led by a board of directors of governors and business leaders.

The National Science Teachers Association (NSTA) is the largest organization in the world composed of professional science educators. NSTA's current membership of 60,000 includes science teachers, science supervisors, administrators, scientists, business and industry representatives, and others involved in and committed to promoting excellence and innovation in science teaching and learning for all.

NSTA's Role

As a partner in the comprehensive process, NSTA has made significant contributions in a number of ways. In 2008 NSTA initiated the program *Science Anchors*, which in many ways spawned the *Framework* and the *NGSS*. From the cadre of science educators who worked on *Anchors* and others, NSTA recommended a number of teachers, professors, and science educators to NRC and Achieve Inc. as writers of the *Framework* and the *NGSS*.

NSTA, in its role as "critical friend" provided two substantive reviews of the *Framework* to NRC; one in the summer of 2010 for the draft *Framework* and the second to Achieve when the *Framework* was released in July 2011. During the development of the *NGSS*, NSTA conducted reviews of three private and two public drafts. The reviews resulted in a detailed analysis and a critique of aspects of the document and a series of recommendations to the writers.

As a partner in the overall process, a major and ongoing role of NSTA is to support its members, the science education community, school administrators, and policy makers in the implementation of the new standards.

CHAPTER 3
INTRODUCTION TO THE NGSS

The Anatomy of a Standard

The *Next Generation Science Standards (NGSS)* consists of a series of standards for grades kindergarten through 12 such as the example standard for grade 2 that is shown in Figure 3.1 (p. 12). The standard page (in the higher grade levels there will be as many as three pages) consists of a title and code, performance expectation, foundation box, and connection box. The document does not, however, precisely define what components of the page constitute the "standard." The reason is, each state that adopts the *NGSS* will need flexibility to assemble the components in a way that meets the needs of that state. The identification of the components is detailed in Figure 3.2 (p. 14), but a short overview is a helpful place to start.

Title and Code

The top of the page contains a code and title that describe the content of the standard. The grade level is designated by the first number—"2" in the example in Figure 3.1—followed by a code—"PS1"—which stands for the first set of ideas in Physical Science. For middle and high school, you will find MS and HS rather than a number representing grade level.

Performance Expectations

The performance expectations describe what a student is expected to be able to do at the completion of instruction. The statement of performance includes a phrase for each of the three "dimensions"—a practice, a disciplinary core idea, and a crosscutting concept—that the *Framework* specifies must be integrated in the performance expectation. These are identified by the color that corresponds to the appropriate dimension in the foundation box below it. Performance expectations are intended to guide the development of assessments, but they are not the assessment as such. They are not instructional strategies or instructional objectives, but they should influence and guide instruction. The listed order of performance expectations does not imply a preferred order for instruction. Note that most of the performance expectations also contain a clarification statement and an assessment boundary statement to provide clarity to the performance expectations and guidance to the scope of the expectations, respectively.

Foundation Box

The foundation box, which follows next on the page and actually comprises three colored columns, contains the learning goals that students should achieve and that will be assessed using the performance expectations. The three parts of the foundation box are

1. science and engineering practices (blue),
2. disciplinary core ideas (orange), and
3. crosscutting concepts (green).

The material contained in the foundation box is taken directly from the respective chapters in the *Framework*. The foundation box also contains learning goals identified as

FIGURE 3.1. A sample standard from the *NGSS* for grade 2

2-PS1 Matter and its Interactions

2-PS1 Matter and its Interactions
Students who demonstrate understanding can:
2-PS1-1. Plan and conduct an investigation to describe and classify different kinds of materials by their observable properties. [Clarification Statement: Observations could include color, texture, hardness, and flexibility. Patterns could include the similar properties that different materials share.]
2-PS1-2. Analyze data obtained from testing different materials to determine which materials have the properties that are best suited for an intended purpose.* [Clarification Statement: Examples of properties could include, strength, flexibility, hardness, texture, and absorbency.] [Assessment Boundary: Assessment of quantitative measurements is limited to length.]
2-PS1-3. Make observations to construct an evidence-based account of how an object made of a small set of pieces can be disassembled and made into a new object. [Clarification Statement: Examples of pieces could include blocks, building bricks, or other assorted small objects.]
2-PS1-4. Construct an argument with evidence that some changes caused by heating or cooling can be reversed and some cannot. [Clarification Statement: Examples of reversible changes could include materials such as water and butter at different temperatures. Examples of irreversible changes could include cooking an egg, freezing a plant leaf, and heating paper.]
The performance expectations above were developed using the following elements from the NRC document *A Framework for K–12 Science Education*:

Science and Engineering Practices	Disciplinary Core Ideas	Crosscutting Concepts
Planning and Carrying Out Investigations Planning and carrying out investigations to answer questions or test solutions to problems in K–2 builds on prior experiences and progresses to simple investigations, based on fair tests, which provide data to support explanations or design solutions. • Plan and conduct an investigation collaboratively to produce data to serve as the basis for evidence to answer a question. (2-PS1-1) **Analyzing and Interpreting Data** Analyzing data in K–2 builds on prior experiences and progresses to collecting, recording, and sharing observations. • Analyze data from tests of an object or tool to determine if it works as intended. (2-PS1-2) **Constructing Explanations and Designing Solutions** Constructing explanations and designing solutions in K–2 builds on prior experiences and progresses to the use of evidence and ideas in constructing evidence-based accounts of natural phenomena and designing solutions. • Make observations (firsthand or from media) to construct an evidence-based account for natural phenomena. (2-PS1-3) **Engaging in Argument from Evidence** Engaging in argument from evidence in K–2 builds on prior experiences and progresses to comparing ideas and representations about the natural and designed world(s). • Construct an argument with evidence to support a claim. (2-PS1-4) *Connections to Nature of Science* **Science Models, Laws, Mechanisms, and Theories Explain Natural Phenomena** • Science searches for cause and effect relationships to explain natural events. (2-PS1-4)	**PS1.A: Structure and Properties of Matter** • Different kinds of matter exist and many of them can be either solid or liquid, depending on temperature. Matter can be described and classified by its observable properties. (2-PS1-1) • Different properties are suited to different purposes. (2-PS1-2),(2-PS1-3) • A great variety of objects can be built up from a small set of pieces. (2-PS1-3) **PS1.B: Chemical Reactions** Heating or cooling a substance may cause changes that can be observed. Sometimes these changes are reversible, and sometimes they are not. (2-PS1-4)	**Patterns** • Patterns in the natural and human designed world can be observed. (2-PS1-1) **Cause and Effect** • Events have causes that generate observable patterns. (2-PS1-4) • Simple tests can be designed to gather evidence to support or refute student ideas about causes. (2-PS1-2) **Energy and Matter** • Objects may break into smaller pieces and be put together into larger pieces, or change shapes. (2-PS1-3) --- *Connections to Engineering, Technology, and Applications of Science* **Influence of Engineering, Technology, and Science, on Society and the Natural World** • Every human-made product is designed by applying some knowledge of the natural world and is built by using natural materials. (2-PS1-2)

Connections to other DCIs in this grade-level: will be available on or before April 26, 2013.
Articulation of DCIs across grade-levels: will be available on or before April 26, 2013.
Common Core State Standards Connections: will be available on or before April 26, 2013.
ELA/Literacy –
Mathematics –

*The performance expectations marked with an asterisk integrate traditional science content with engineering through a Practice or Disciplinary Core Idea.

Source: Achieve. 2013. *Next generation science standards.* www.nextgenscience.org/next-generation-science-standards.

1. connections to engineering, technology, and applications of science (found in the green crosscutting area); and
2. connections to the nature of science (found in the practices area in this example but also can be found in the crosscutting concepts).

These supplemental goals are related to the other material in the foundation box and are intended to guide instruction, but the outcomes are not included in the performance expectations.

Appendix H, Nature of Science, and Appendix J, Science, Technology, Society, and the Environment, in the *NGSS* document contain more useful information about this material.

> Although the *NGSS* does not define a standard, NSTA considers a standard to be the performance expectations and foundation box associated with a core idea at a given grade level or band.

Connection Box

To support instruction and development of instructional material, a connection box appears immediately below the foundation box. This box

1. identifies connections to other disciplinary core ideas at this grade level that are relevant to the standard (it contains the code for standards in other core ideas);
2. identifies the articulation of disciplinary core ideas across grade levels (it contains the codes of other standards in both prior and subsequent grade levels); and
3. identifies connections to the *Common Core State Standards (CCSS)* in mathematics and in English language arts and literacy that align to this standard (note that each *CCSS* standard is followed by a reference to a performance expectation).

What Are Standards?

Although the above information describes the anatomy or contents of a standard and its supporting connection box, it is important to have a more fundamental idea—independent of the *NGSS*—as to what standards are and how they relate to instruction, curriculum, and assessment. This understanding is important so that the role and purpose of the *NGSS* is understood and misconceptions do not arise during its use.

- Standards shed only partial light on how instruction should be conducted to meet the goals in the standards. The front matter of the *NGSS* discusses instruction but does not specify the exact nature of it for any given standard. The practices specified in a performance expectation may suggest a type of activity or behavior, but the practices do not define the type or nature of instruction.
- Standards are for all students. They can be considered an achievement level that all students should attain but not an average level of attainment. Performance for any given student or group of students may be higher as a result of instruction and expectations

FIGURE 3.2. Inside the *NGSS* Box, a key to the various elements of a page in the *NGSS*

designed to reach these higher goals. See a discussion of this topic in Appendix D, All Standards, All Students, in the *NGSS* document.
- Standards are not a plan for curriculum or a curriculum framework. Individual standards may specify learning goals for a given grade or grade band, but the order or arrangement of instruction and the learning goals within a grade level or band is not specified.
- Standards have strong implications for the professional preparation of teachers at the preservice level and the ongoing professional development of practicing teachers, but they do not specify the nature or extent of the preparation.

In rather straightforward terms, the *NGSS* has only two specific purposes beyond its broad vision for science education, namely (1) to describe the essential learning goals, and (2) to describe how those goals will be assessed at each grade level or band. The rest—instruction, instructional materials, assessments, curriculum, professional development, and the university preparation of teachers—is up to the science education community.

CHAPTER 4
MOVING FROM THE NGSS TO INSTRUCTION

> The *Framework* is designed to help realize a vision for education in the sciences and engineering in which students, over multiple years of school, actively engage in scientific and engineering practices and apply crosscutting concepts to deepen their understanding of the core ideas in these fields. (*Framework*, pp. 8–9)

Introduction to the Design Process

There is no better way to begin the process of becoming acquainted with the substance of the *Next Generation Science Standards (NGSS)* and meeting this new vision than to use the document to focus on student learning by planning instruction and instructional materials. To do so, you must understand the *NGSS* architecture, interpret the meaning of the various components of standards and their relationship to one another, and then translate them into instruction. The previous chapter described the structure or architecture of the *NGSS*. This chapter outlines a dynamic, active process for understanding how each component of a standard interacts with the other components by developing an instructional sequence starting with one or possibly two performance expectations. It is a roadmap but not a fully detailed manual for developing instruction. Chapter 5 describes a more extensive group process to study the entire document with the goal of a comprehensive curriculum design in mind.

It is generally accepted that designing instruction or instructional material begins with the selection of the endpoint, or desired results of the instruction, and working backward through an instructional sequence to the beginning knowledge students have coming into the instruction. The description of such a process has been well documented by Wiggins and McTighe in *Understanding by Design* (1998). The process outlined here will draw modestly from the concepts in that backward design process. The *NGSS* is an excellent starting point since its purpose is to outline the outcome in terms of learning goals based on science and engineering practices, disciplinary core ideas, crosscutting concepts and the performances expectations that are derived from them at the various levels of progress from kindergarten through high school.

Creating a Description of the Desired Results

Begin by establishing the desired results of instruction or the learning goals associated with one performance expectation. In reality, the goals for two or possibly three closely related performance expectations could be combined in a learning sequence, but for the purposes of this illustration, using only one will simplify the description and understanding of the process. (For the purposes of this activity, select a performance expectation that does not include an engineering component as indicated with an asterisk. We will deal with these performance expectations more thoroughly in the next chapter.)

Keep in mind that the purpose of the performance expectation is to describe the performance expected of the student during the assessment process **after** instruction is complete. **It is not a learning objective as such and certainly is not the description of an instructional**

strategy. Since it is a description of assessment, we need to be reminded that the assessment of a learning goal is by its nature limited to one context and one practice, one crosscutting concept, and possibly only a portion of one disciplinary core idea. It is only one of many methods of assessing the three dimensions referenced in the performance expectation. There are many other possible ways to assess the same set of dimensions; the performance expectation provided is only one possible way. In addition, it is only a sampling of all the possible learning goals that could be considered for the dimensions represented in the performance expectation. Keep in mind that your school or your district may have additional expectations for students who have mastered the learning goals in the *NGSS*.

The performance expectation you select will link or connect you to at least one practice, a disciplinary core idea, and a crosscutting concept in the foundation box.

It may be useful to refer to Figure 3.2, Inside the *NGSS* Box, on page 14, which defines each element on the page. Note how the content in the foundation box is keyed to one or more performance expectations.

Use the content in the three dimensions in the foundation box to create a description of the desired results, or endpoint, of the instructional sequence called for in the backward design process. The performance expectation in the standard later becomes a description of **one** method of assessing the students' proficiency in meeting those results.

Developing Activities From a Standard

In most cases, the activities in the instructional sequence should integrate the three dimensions of practices, disciplinary core ideas, and crosscutting concepts to be consistent with the charge from the *Framework* and the nature of instruction necessary to create success in the integrated outcome described in the performance expectations. Developing the activities in the instructional sequence requires a careful examination and use of the three dimensions: science and engineering practices, the disciplinary core idea, and the crosscutting concept. To do this, read the material in the foundation box that is coded for the performance expectation you selected. In addition to the material in each dimension, go to the description of the practice, disciplinary core idea, and crosscutting concept in the *Framework*. Although the three dimensions will be integrated, we will consider them individually here in the early planning process here.

Download the *Framework* for free at www.nap.edu or purchase a hard copy from NSTA at www.nsta.org/store. NSTA has produced a number of resources to help science educators better understand the *Framework* and its dimensions, including *The NSTA Reader's Guide to A Framework for K–12 Science Education* and a number of articles published in NSTA member journals. These resources are available at www.nsta.org/ngss. *The Reader's Guide* and many of the journal articles have also been bundled into one volume, *The NSTA Reader's Guide to A Framework for K–12 Science Education, Expanded Edition*, available at www.nsta.org/store.

Disciplinary Core Idea

You will find a one- to two-page essay in *Framework* Chapters 5, 6, 7, and 8 discussing each disciplinary core idea and four endpoints for grades 2, 5, 8, and 12. As you read the description and endpoints and create the activities in the learning sequence consider the following questions:

- What are some commonly held student ideas (both troublesome and helpful) about this topic? How could instruction build on the helpful ones and address the troublesome ones?
- What prior ideas or concepts do students need to learn to understand this core idea? What level of abstraction is expected of students?
- What are some phenomena and experiences that could provide observational or experimental evidence that that targeted core idea is an accurate description of the natural world?
- What representations or media would be helpful for students to use in making sense of core ideas?

Science and Engineering Practices

Framework Chapter 3 provides essays for each science and engineering practice and a discussion of progression along the grade levels, but no specific endpoints other than grade 12. The practices in the foundation box were developed by the *NGSS* writers for each grade level or band.

- The practice in the performance expectation is the one to be used in the assessment process. Many other practices from the full set of eight will need to be included in the instructional sequence. Which ones and in what order will you use them?
- How will each practice be used to develop an understanding of the disciplinary core idea?
- What practices could students engage in to explore phenomena and/or representations?

Crosscutting Concepts

The same format is available for the crosscutting concepts in Chapter 4 of the *Framework*. As with the practices, the *NGSS* writers created the endpoints for each practice for grades 2, 5, 8, and 12 that appear in the foundation box. As you create the instructional sequence consider the following questions:

- How will the crosscutting concepts indicated in the performance expectation support the understanding of the core idea?
- Are there other crosscutting concepts that could also support learning the core idea?

Before leaving the overview of the material in the standard, locate the last connection box at the bottom of the page and consider what connections to the *Common Core State Standards (CCSS)* in mathematics and English language arts could be emphasized as students engage in the instructional sequence. More on these connections in the next chapter.

Putting It All Together

The above review of the three dimensions coupled with the embedded questions and comments should provide the ingredients from which the activities in the instructional sequence can be created. The next task is the use of an integrated instructional model to actually create the sequence. One of the most widely used, well-described, and researched models for designing an instructional sequence is the BSCS 5E Model developed by Rodger Bybee. Bybee uses this model to develop an instructional sequence starting with the standards in the *NGSS* in a forthcoming

NSTA publication, *Translating the NGSS for Classroom Instruction*. This would be an excellent resource for guidance in using the 5E Model to develop the integrated instructional sequence.

Creating an Assessment

The last step, which is also included in the 5E Model, is a return to the first step, where a description of the desired results was created. The summative assessment is a refinement, if necessary, of the previous assessment constructed as an early step in the backward design process. While refining, consider the following questions:

- Is the assessment procedure or task aligned and consistent with the performance expectation?
- Are all three dimensions assessed?
- What intermediate or formative assessment can be created to check student progress and modify instruction during the instructional sequence?

Conclusion

This simplified process is only the beginning of a more elaborate and often reiterative process of writing, reviewing, piloting, and rewriting that goes into the development of instructional sequences and quality instructional materials. Nevertheless, it can be a starting point for interpreting a portion of a standard and clarifying the fundamental nature of how to use the various components of the standards in the process. The next chapter provides more details on how to organize a group to review the entire *NGSS* and think more deeply about designing instruction, developing the total K–12 curriculum, and resolving professional development issues.

The entire process of developing an instructional sequence based on a standard is summarized with the sequence of activities in Figure 4.1.

FIGURE 4.1. A summary of the process for moving from the *NGSS* to instruction

1. Select a performance expectation.
2. Read the performance expectation, clarification statement, and assessment boundary.
3. Read the applicable disciplinary core idea in the foundation box.
4. Read material in the *Framework* for the disciplinary core idea cited.
5. Read the science and engineering practices in the foundation box related to the performance expectation.
6. Read the material in the *Framework* for these practices.
7. Read the crosscutting concept in the foundation box associated with the performance expectation.
8. Read the material in the *Framework* for this crosscutting concept.
9. Create one or more descriptions of the desired results or learning goals for the instruction integrating the three dimensions in the foundation box.
10. Determine the acceptable evidence for the assessment of the desired results. Draft the summative assessment process for the learning goal.
11. Create the learning sequence using the BSCS 5E Model.
12. Create the summative assessment and check its alignment with the performance expectation.

CHAPTER 5
A GUIDE FOR LEADING A STUDY GROUP ON THE TOTAL CURRICULUM

Chapter 4 provided a procedure to acquaint you with the substance and structure of a portion of a single standard in the *Next Generation Science Standards (NGSS)* by taking you through an exercise of translating a performance expectation and the associated content in the foundation box into a limited instructional sequence. It was not an exhaustive process of developing a "ready to use" learning sequence but rather a means to help you understand how the components of a standard interact and can be used to create an instructional sequence of activities. It could be completed as an individual or a team with each member using a different performance expectation and comparing results at the conclusion of the work.

The procedure outlined in this chapter is the next step in understanding the full array of what is in the *NGSS* and will prepare your team for the more comprehensive step of examining what is in the entire document, considering the learning progressions, and beginning the development of a multiyear (K–12, K–5, 6–8, or 9–12) curriculum. The questions in this chapter are meant to assist you in thinking about how standards or parts of standards are grouped together and sequenced and where they are located in the curriculum. The questions help you continue to think about the instructional strategies and assessments that you need. The process requires a team because it is a comprehensive study of the entire *NGSS* or any subset of the document that a team selects.

Using a group to study the *NGSS* and plan its use or implementation has a number of advantages. The *NGSS* is a large and complex document that spans 13 grade levels and four major content domains, requiring a leadership team with a wide variety of experience and expertise. The task of developing instructional materials and providing professional development requires a team of several individuals in most school or district situations.

During the development of the *NGSS*, NSTA created a guide to help educators plan and facilitate a group study of the various draft standards. That guide has been modified here to foster understanding and interpretation of the *NGSS* by a team of people working together as a study group. This group work is an important part of the process of using the standards to guide the development of instructional materials and professional development.

Organizing a Group

Your first steps in organizing a team are determining the scope of study and the number of participants. There needs to be a match between the scope of the study and the group size.

In some cases, the membership of the group is already predetermined and the group leader must set the scope of the study. For example, a group of four high school biology teachers may have decided to work together to study the standards. Since the membership of this group is predetermined, they would then use that information to decide on scope, which would likely be to focus on the life science standards in high school. In other cases, the scope of study may be predetermined and the group leader then needs to decide the membership of the group. For example, a district curriculum coordinator has been asked to form a team to review the standards and develop an implementation plan for grades K–8. Because the scope of task has been set, the curriculum coordinator would be sure to have a group that included a mix of elementary and middle school teachers with expertise in life, Earth, and physical sciences.

A study group may be designed to meet just a few times (or even just once), or it may be designed to meet regularly over the course of a school year (or even longer).

Based on our experience of reviewing the standards during their development, a team of at least 16 people is necessary to conduct a comprehensive review of the overall scope of the standards. Figures 5.1 and 5.2 provide two schemes for organizing and selecting members of the study. Plan A involves a minimum of two people to study the K–2 standards; two people to study the grades 3–5 standards; and two people each to study the middle school and high school life sciences, physical sciences, and Earth and space sciences. Plan B involves a minimum of two people each to study the elementary life sciences, physical sciences, and Earth and space sciences. Engineering and the nature of science will be studied as a second round with the three disciplines. This would be repeated for middle and high school, for a minimum of 18 reviewers. Regardless of the size of the team, the emphasis should be on depth rather than breadth. The *NGSS* is a very complex document that is best understood by digging deeply into it. It is much more important to have an in-depth exploration of a few sections of the document than a limited look at many sections.

FIGURE 5.1. Size and scope of study group Plan A

	Life Science	Physical Science	Earth and Space Science
K–2	2–5 team members		
3–5	2–5 team members		
6–8	2–5 team members	2–5 team members	2–5 team members
9–12	2–5 team members	2–5 team members	2–5 team members

FIGURE 5.2. Size and scope of study group Plan B

	Life Science	Physical Science	Earth and Space Science
K–2	2–5 team members	2–5 team members	2–5 team members
3–5			
6–8	2–5 team members	2–5 team members	2–5 team members
9–12	2–5 team members	2–5 team members	2–5 team members

Becoming Familiar With the *Framework*

As noted earlier, the most important aid in interpreting the standards is reviewing its parent document, *A Framework for K–12 Science Education* (NRC 2012). The *Framework* describes the major practices, crosscutting concepts, and disciplinary core ideas that all students should be familiar with by the end of high school, and how these practices, concepts, and ideas should be developed across the grade levels. Much of the material in the *NGSS* foundation box comes verbatim from the *Framework*, and the *Framework* contains additional background information about the material in the foundation box. Many of the suggested topics for the study groups listed below include reading material from the *Framework*.

Examining a Page of the Standards

Study Group Discussion Topics

One of the most useful activities for a group of two is to carefully examine a page of the standards using the following list of discussion topics. The suggested activity can be done many times by focusing on a different page in the standards each time.

The *NGSS* document is organized in two separate configurations: (1) by topic, and (2) by disciplinary core idea. States and districts can choose to use either one or to design an alternative way of organizing them. Before you begin, determine which version of the standards (by topic or by core idea) you will use for your team. If your state has adopted the *NGSS*, you might check and see if it has a particular way of organizing the standards and use that configuration.

Disciplinary Core Ideas

Start by examining the disciplinary core idea followed by a close look at the practices and crosscutting concepts associated with the performance expectation. Then return to the performance expectation to review a description of how the content in the foundation boxes will be assessed. For each performance expectation, note the related disciplinary core ideas in the foundation box and consider the following questions:

- Is there one key idea or several separate ones in each disciplinary core idea? If there is more than one, clarify each key idea separately.
- What is the relationship between the overall set of core ideas? What contexts could be used to address them coherently in an instructional unit? Would the coherence of the set be improved by moving one of the core ideas to a different set or transferring additional core ideas to this set?
- Is the overall set of core ideas too much for one instructional unit? Is it too little? Is it fine as is or should it be broken up into smaller groups of core ideas? (Or merged with other sets to make a larger group?)
- How do the disciplinary core ideas referenced in each performance expectation at this grade level build on those from earlier grades? To answer this question, it will be helpful to refer to the connection box for articulation of disciplinary core ideas across grade levels,
- How does the grade level placement of this content compare with where it is addressed in your existing curriculum?

It may be useful to refer to Figure 3.2, Inside the *NGSS* Box, on page 14, which defines each element on the page. Note how the content in the foundation box is keyed to one or more performance expectations.

Science and Engineering Practices
- How do the practices in the foundation box support an understanding of the disciplinary core idea?
- What specific techniques or strategies will students be expected to use to engage in this practice?
- Is the practice being used in a scientific context or an engineering context? What difference (if any) does that make in how students engage in the practice?

Crosscutting Concepts
- How do the crosscutting concepts in the foundation box support understanding of the associated disciplinary core idea?
- Will the crosscutting concepts assist or support in learning the disciplinary core idea or will the process work in reverse? Namely, as the understanding of the disciplinary core ideas develops, will it aid in understanding the broader crosscutting concept?

Performance Expectations
- What do students need to be able to know and do to demonstrate that they have achieved the performance expectation?
- What is an example of a specific task(s) that students could do to demonstrate their achievement?
- In what ways does the clarification statement help define what students must know and be able to do?
- In what ways does the assessment boundary define what can be assessed?
- What will a student's successful achievement of a performance expectation tell you about his or her understanding of the targeted core idea(s)?
- What will a successful performance of the specified practice look like?

Engineering, Technology, and Applications of Science
Engineering shows up in several different ways in the *NGSS*. To understand the multiple aspects related to engineering you should first read Appendix I, Engineering Design in the *NGSS*. As Appendix I indicates, engineering is present as (1) engineering design; and (2) links among engineering, technology, science, and society. These two aspects appear in a standard in an engineering practice, disciplinary core idea, or crosscutting concept. The performance expectations that incorporate any of these three dimensions are marked with an asterisk.

Ask participants—working individually or in teams of two—to return to the set of standards on the page they were originally assigned and review the performance expectations indicated with an asterisk.

- Carefully note how the performance expectation incorporates an aspect of engineering. Most of the links among engineering, technology, science, and society are included in the crosscutting concepts but are not represented in the performance expectations.

- If the performance expectation also involves core ideas in life science, Earth science, or physical science, how can instruction address this material in conjunction with the engineering material?
- Does an understanding of the core idea in the discipline need to be achieved before the engineering performance is accomplished, or will the engineering dimension facilitate the learning of the disciplinary core idea?
- What are some examples of how each of the eight practices can be used in an engineering context? (Try finding an example performance expectation in the standards for each practice and also try to come up with some of your own examples.)

Nature of Science

To support the learning of the nature of science, various aspects have been placed in the practices and the crosscutting concepts. A discussion of why and how this was done is found in Appendix H, The Nature of Science, in the *Next Generation Science Standards*. A matrix of the complete learning progress of the nature of science can be found in this appendix.

Although these aspects of the nature of science appear in the foundation box, they are not incorporated in the performance expectations.

Ask participants—working individually or in teams of two—to return to the set of standards they were originally assigned and review the practices and crosscutting concepts where nature of science ideas have been integrated following the same procedure as above. You can then use these discussion questions to engage in a conversation about the connections to the nature of science:

- Do you recommend that the nature of science be included in your school or district as an explicit part of the curriculum? If so, will you recommend that it be assessed?
- How would understanding this connection help students see the strengths and limitations of science?
- To what extent do your current methods of instruction provide opportunities for students to learn the nature of science? Could they be improved?
- How could you assess a student's understanding of the nature of science?

Connections to Other Disciplinary Core Ideas at This Grade Level

This is the first connection box immediately below the foundation box.

- Can or should any of these disciplinary core ideas be grouped with the ones in this standard in a cohesive way? Is the reverse possible? This may depend on whether you are creating interdisciplinary units at the same grade level.
- How can instruction be designed so that students note the connections between the core ideas?

Articulation of Disciplinary Core Ideas Across Grade Levels
- Examine the standards in the same topic at earlier grade levels. Do they provide an adequate prior knowledge for the core ideas in the standard being reviewed?

- Examine the standards in the same topic at later grade levels or bands. Does the standard at this level provide an adequate prior knowledge for the core ideas in the later standards?

Connections to Common Core State Standards in Mathematics
- Should students have achieved these mathematics standards to engage in the learning of the science, or could the mathematics and science be learned together?
- In what ways do the referenced mathematics standards help clarify the science performance expectations description of what students are expected to know and be able to do?
- Can any of the science core ideas be included as examples in the mathematics instruction?

Connections to Common Core State Standards in English Language Arts
- Should students have already achieved these English language arts standards to engage in the learning of the science, or could the English language arts and science be learned together?
- In what ways do the referenced English language arts standards help clarify the performance expectations description of what students are expected to know and be able to do?
- Can any of the science core ideas be included as examples in the English language arts instruction?

After the notes and reports from the different grade levels, grade bands, or disciplines have been discussed and assembled into a report or summary, they can be reviewed as a resource while a variety of decisions or plans are being made related to the implementation of the *NGSS*. These plans include

- a determination of the type and amount of professional development required;
- the degree to which the structure of the K–12 curriculum needs to be redesigned;
- the amount of new or revised instructional materials needed;
- a sense of the time and cost necessary to effect the implementation; and
- next steps in planning and carrying out the implementation.

The extent and scheduling of these decisions depends on when and to what degree the *NGSS* will be implemented. The next chapter provides an overview of the issues involved in a comprehensive implementation of the new standards.

CHAPTER 6
IMPLEMENTATION

Introduction

We concluded Chapter 3 with the position that the *Next Generation Science Standards (NGSS)* has two specific purposes beyond its broad vision for teaching and learning, namely (1) to describe the essential learning goals in the foundation boxes, and (2) to describe how those goals will be assessed at each grade level or band in the performance expectations. The vision of the *Framework* and the *NGSS* will be realized only if significant changes in instruction, instructional materials, assessments, curriculum, and professional development are implemented at the state and local levels. This chapter briefly discusses the planning and work that needs to go into "implementing" the *NGSS* by effecting the changes called for in the documents. There is much work to be done at the college and university level in the preparation of teachers for their role as partners with state and local education agencies, but describing work that is unique to the university level is beyond the scope of this guide.

How Much Change Will Be Needed?

The answer to that question, obviously, depends on your existing practice. The degree of change and the nature of the change will vary considerably from teacher to teacher, from school to school, and from district to district. But in this guide, as a starting point, I describe the best estimate of the changes called for and the effort and resources needed using a group of teachers who have done a reasonably comprehensive job of implementing the *National Science Education Standards* or the *Benchmarks for Science Literacy*. There will be a number of exceptions or variations to this starting point. However, the reform efforts needed for teachers and schools at "ground zero" to provide a quality science program is far beyond the scope of this guide.

Start with a close reading of a very informative *NGSS* document, Appendix A, Conceptual Shifts, in the *Next Generation Science Standards*. Consider it a broad statement of the programmatic goals of the *NGSS* rather than as a description of the shifts in the document. We will reference this useful appendix as we discuss the changes called for in the *NGSS*.

Achieve and the U.S. Education Delivery Institute have produced a publication to assist states in their adoption activities, *Next Generation Science Standards: Adoption and Implementation Workbook,* that you may find very helpful.

What Changes Are Needed and What Is Required to Make Them?

Integration of practices, disciplinary core ideas, and crosscutting concepts. Probably the most significant innovation in the *NGSS* is the integration of the three dimensions in the performance expectations. The first shift detailed in Appendix A discusses this idea, which was first introduced in the *Framework,* then applied in the *NGSS*. The integration applies not just to how assessments will be designed, but also to the idea that science should be learned through the "intertwining" of the three dimensions in K–12 classrooms.

Practices as assessed outcomes and instructional methodology. Many teachers already include inquiry strategies in their instruction but rarely are they considered to be outcomes that are assessed. For these teachers, the shift means making sure the complete set of science and engineering practices are included in their overall instruction and developing means to assess a student's ability to carry out the practices independently. This will, in most cases, require new instructional strategies and materials. Review the information in Appendix F, Science and Engineering Practices, in the *NGSS*. This topic is discussed further later in the chapter.

To better understand the shift required for teachers already using inquiry as defined in the *NSES*, consider the similarities between inquiry "abilities," as they were called in the *NSES*, and the new practices described in the *NGSS*, as illustrated in Figure 6.1.

FIGURE 6.1. A comparison of Abilities of Inquiry from the *NSES* with science and engineering practices from the *NGSS* (The order has been changed in the Abilities of Inquiry to correspond to the most relevant practice.)

Science and Engineering Practices NGSS	Inquiry (Abilities) NSES Grades 5–8
Asking questions and defining problems	Identify questions and concepts that guide scientific investigations
Developing and using models	Develop descriptions, scientific explanations, and models using evidence
Planning and carrying our investigations	Design and conduct scientific investigations
Analyzing and interpreting data	Use appropriate tools to gather, analyze, and interpret data
Using mathematics and computational thinking	Use mathematics in all aspects of inquiry
Constructing explanations and designing solutions	Recognize and analyze alternative explanations and predictions
Engaging in arguments from evidence	Think critically and logically to make relationships between explanations and evidence
Obtaining, evaluating, and communicating information	Communicate a scientific procedure and explanations

Integrating crosscutting concepts. The crosscutting concepts will require modification of instruction and instructional materials for many teachers. Although the idea of common themes (AAAS 1993) or unifying concepts and principles (NRC 1996) have been present for many years, it has been rarely emphasized or assessed. The similarities and differences in common themes and unifying concepts and principles when compared to crosscutting concepts is illustrated in Figure 6.2. Implementation of the *NGSS* will require the explicit addition of the crosscutting concepts in most lessons. Formative and summative assessment procedures and methods will need to be developed.

FIGURE 6.2. Comparison of crosscutting concepts (*NGSS*) with unifying concepts and processes (*NSES*) and common themes (*Benchmarks*)

NGSS Crosscutting Concepts	NSES Unifying Concepts and Processes	Benchmarks Common Themes
Patterns	Systems order and organization	Models
Cause and effect	Evidence, models, and explanations	
Scale proportion and quantity	Change, constancy, and measurement	Scale
Systems and system models	Evolution and equilibrium	Systems
Energy and matter: Flow, cycles, and conservation		Constancy and change
Structure and function	Form and function	
Stability and change		

When crosscutting concepts are included in instruction, their role needs to be considered: Are they contributing to the understanding of the disciplinary core idea, or is the understanding of the crosscutting idea being enhanced by the disciplinary core idea? In the early elementary grades, the crosscutting concepts are only beginning to emerge in students' minds and usually in very specific contexts, requiring instruction to be explicit in identifying the concept and illustrating how it is useful in other contexts. As an example, primary level students may see the pattern in the arrangement of a few blocks of different color and shapes but not realize that there are patterns in the behavior of small organisms such as crickets or mealworms. The instructor directly pointing out the use of the crosscutting concepts will help students develop the practice of identifying patterns in later experiences. In contrast to this early instruction, high school students who have experienced the use of the crosscutting concepts for several years will be tuned in to looking for patterns in a wide variety of phenomena.

Review the information in Appendix G, Crosscutting Concepts, in the *NGSS*.

Making the Changes

Making Changes in Instructional Materials

Designing, adapting, or adopting instructional materials is a logical early step in the change process since the materials teachers use are a major support for their instruction and are an important ingredient in designing the school or district curriculum. Rodger Bybee developed a scheme for presenting the practices, crosscutting concepts, and disciplinary core ideas in the activities of an instructional sequence by identifying each dimension as foreground or background and by emphasizing each dimension to a greater or lesser extent in a well-planned system. Figure 6.3 illustrates one example using the BSCS 5E model for an instructional sequence that indicates the foreground or background emphasis of each dimension. More details on how to develop instructional materials for an integrated instructional sequence based on the *NGSS* can be found in Bybee's forthcoming book, *Next Generation Science Standards and the Life Sciences* (2013).

Ultimately, a variety of instructional materials based on the *NGSS* will be developed by a variety of sources such as state and local education authorities, nonprofit science education organizations, and commercial publishers. Regardless of the source, teachers and schools will need criteria to establish guidelines for development of new instructional materials or for evaluation of completed and available products. NSTA is developing a rubric in cooperation with Achieve Inc. to judge the degree of alignment of materials with the *NGSS*. That rubric will be available at *www.nsta.org/ngss*.

Another excellent resource to evaluate the alignment is AIM (Assessing Instructional Materials), developed by the K–12 Alliance at WestEd and adapted by BSCS (Short 2006). The AIM process uses evidence to analyze and select instructional materials in what is as much a professional development process as it is a material selection process. It is a collaborative endeavor in which the participants develop the criteria to be used before collecting evidence from the material to score the materials and make a decision. Suggestions for the criteria are based in part on the inquiry standards in the *NSES* but could easily be modified to include the practices and crosscutting concepts from the *NGSS*.

Other noteworthy selection processes have been developed by NRC (1999) and AAAS—Project 2061 (Kesidou and Roseman 2002).

Making the Changes in Curriculum

As noted earlier, the *NGSS* does not determine the curriculum that needs to be developed to assure that all students are able to meet the expectations of the standards. Because the *NGSS* is anchored in the idea of learning progressions wherein concepts build coherently from grades K–12, it does have a significant influence on the arrangements of the disciplinary core ideas and the related performance expectations. Fortunately, Achieve Inc. attached a very helpful Appendix K to the standards, which suggests three possible "course maps" and the procedure used to create the maps. These maps are an excellent starting point for developing state and local district curriculum.

FIGURE 6.3. A 5E instructional sequence with different emphasis on the dimensions (Bybee, forthcoming)

Activity	Practices	Crosscutting Concepts	Disciplinary Core Ideas
ENGAGE • What teacher does • What students do			
EXPLORE • What teacher does • What students do			
EXPLAIN • What teacher does • What students do			
ELABORATE • What teacher does • What students do			
EVALUATE • What teacher does • What students do			

Using Professional Development to Modify Instruction

Surveys that NSTA conducted with its members that asked about their perceived needs for professional development (NSTA 2013b) have indicated a strong desire for offerings that help them do the following:

- Interpret the standards
- Understand the relationship between performance expectations and instruction
- Design instructional strategies
- Address the needs of students with different levels of ability
- Incorporate engineering and technology in the curriculum
- Assess whether students have achieved the standards
- Know what putting the *NGSS* into practice looks like in the classroom

The task of designing a comprehensive professional development program is beyond the scope of this *Reader's Guide,* but the outline used here and shown in Figure 6.4 would be useful as a framework for a plan that will help you move from awareness, through interpretation, to the early stages of implementation of the *NGSS.*

> **FIGURE 6.4.** Professional development framework using the plan outlined in this guide
>
> I. The *Framework* and background history
> II. Introduction to the *NGSS*
> III. Moving from the *NGSS* to Instruction
> IV. An in-depth review of the *NGSS* using a study group
> V. Implementation
>   A. Instructional materials: Developing, evaluating, adapting
>   B. Modifying instruction
>
> *Note:* Achieve Inc. has published a document (Achieve and U.S. Education Delivery Institute 2013) for state leaders who plan to adopt and implement the *NGSS* that could be helpful for your team.

The Integration of Engineering May Require Special Attention and Professional Development

If you haven't done so already, read Chapter 8 of the *Framework,* Dimension 3: Disciplinary Core Ideas—Engineering, Technology, and Applications of Science, and then *NGSS* Appendix I, Engineering Design in the *NGSS.* This comprehensive appendix gives a good idea of why and how engineering has been integrated in the other three core ideas in the *NGSS.* Note the definitions of science, engineering, and technology on page 1 of the appendix; these may be somewhat different from what you have used in the past.

One consideration not specified in the appendix is the challenge that performance expectations labeled with an asterisk present in instruction and assessment. Students are expected to have achieved the science learning goal and the associated engineering design goal, in addition to the practices and crosscutting concepts. Knowing the science concepts will be necessary to carry out the problem-solving stages from the engineering design model called for in the performance expectation.

Making the Changes in Assessment

As noted earlier, performance expectations are not assessment items or an assessment process. They describe the performance that should be assessed using any number of media, such as traditional paper and pencil tests, online assessments, practical or hands-on assessments, or a combination of these. The performance expectations should not be considered a description of the *only* means or context for assessment. Classroom instructors and material developers should consider a variety of assessment means, especially the use of formative assessment tasks throughout the instructional sequence. When large-scale assessments are available in the future, these instruments will be expected to use the performance expectation as the guide or framework for the assessments. (At the time this guide was published there were no known plans under way to develop a large-scale assessment. It is anticipated that this will occur but because of the time and money required to develop and pilot such an assessment, it would not be available for a number of years.)

Developing Support

This last step, and one of the most important ones, is to identify and develop the support necessary for the interpretation and implementation of the *NGSS,* as outlined in this guide. Creating awareness, understanding, and implementing cannot be carried without financial and administrative support. Do not overlook the need for time, in addition to the more tangible resources. Some of the early steps described in this guide require only limited written materials and equipment, but financial resources are needed for extensive release time and a qualified facilitator. NSTA surveys (NSTA 2013b) indicate that extensive professional development, instructional materials, and equipment are needed to support classroom teachers.

Regardless of whether your role is a science teacher, a department head, or a school or district curriculum supervisor, it is your responsibility to identify and clearly document the materials, equipment, or activities that require financial support and to present it to the responsible administrator or school board. Support will not happen without initiative—take it! We all have responsibility in making the needs known and getting the required support.

Final Note

As Chapter 1 indicated, this guide is only the beginning of a long and important journey. NSTA has produced and will continue to produce publications, web seminars, conferences, other professional development sessions, position papers, and online communities to support you during the process. Let the association know how it can help you as you begin this very important journey, one critical to your students and our country.

REFERENCES

Achieve Inc. 2013. *Next generation science standards.* www.nextgenscience.org/next-generation-science-standards.

Achieve Inc. and U.S. Education Delivery Institute. 2013. *Next generation science standards: Adoption and implementation workbook.* Washington, DC: Achieve and U.S. Education Delivery Institute.

American Association for the Advancement of Science (AAAS). 1990. *Science for all Americans.* New York: Oxford University Press.

American Association for the Advancement of Science (AAAS). 1993. *Benchmarks for science literacy.* New York: Oxford University Press.

American Association for the Advancement of Science (AAAS). 2001. *Atlas of science literacy.* Washington, DC: AAAS.

Bybee, R. Forthcoming. *Translating the* NGSS *for classroom instruction.* Arlington, VA: NSTA Press.

Carnegie Corporation of New York and Institute for Advanced Study. 2007. *The opportunity equation: Transforming mathematics and science education for citizenship and global economy.* New York: Carnegie Corporation of New York.

Kesidou, S., and J. E. Roseman. 2002. How well do middle school programs measure up: Findings from Project 2061 curriculum review. *Journal of Research in Science Teaching* 39: 522–549.

National Governors Association Center for Best Practices and Council of Chief State School Officers (NGAC and CCSSO). 2010. *Common core state standards.* Washington, DC: NGAC and CCSSO.

National Research Council (NRC). 1996. *National science education standards.* Washington, DC: National Academies Press.

National Research Council (NRC). 1999. *Selecting instructional materials.* Washington, DC: National Academies Press.

National Research Council (NRC). 2012. *A Framework for K–12 science education: Practices, crosscutting concepts, and core ideas.* Washington, DC: National Academies Press.

National Science Teachers Association (NSTA). 2013a. How to lead a study group on the *NGSS.* Arlington, VA: NSTA. *www.nsta.org/about/standardsupdate/resources/HowToConductAStudyGroupOnNGSS.pdf.*

National Science Teachers Association (NSTA). 2013b. Survey of *NGSS* Implementation. Washington, DC: Personal correspondence.

Short, J. B. 2006. *Analyzing standards-based science instructional materials: An opportunity for professional development.* New York: Teachers College Press.

Wiggins, G., and J. McTighe. 1998. *Understanding by design.* Alexandria, VA: Association for Supervision and Curriculum Development.

INDEX

*Page numbers printed in **boldface** type refer to figures.*

A
A Framework for K–12 Science Education, 7–8
   conceptual framework of, 7
   development of, xii, 8
   dimensions of, 7–8
      crosscutting concepts, 7
      disciplinary core ideas, 7–8
      integration of, 8, 27–29, **28, 29**
      science and engineering practices, 7
   funding for, 7
   guiding assumptions and organization of, 7
   obtaining a copy of, 8, 18
   as prerequisite for studying the *NGSS,* 2–3, 7–8
   purpose of, 17
   resources on, 4–5, 18
   study group to become familiar with, 22–23
Achieve Inc., xii, 3, 8, 9, 27, 30
Activities developed from a standard, 18–20, 21
   assessment of, 20
   crosscutting concepts, 19
   disciplinary core idea, 18–19
   integration of dimensions, 18, 19–20
   science and engineering practices, 19
American Association for the Advancement of Science (AAAS), xii, 8, 9
   Project 2061, 9, 30
Anderson, Andy, 6
Assessing Instructional Materials (AIM), 30
Assessment(s)
   of alignment of instructional materials with the *NGSS,* 30
   creation of, 20
   formative, 20, 29, 32
   learning goals for, 11, 13, 15
   making changes in, 32–33
   performance expectations and, 11, 27, 32
   of science and engineering practices, 28
   of student's understanding of the nature of science, 25
Atlas of Science Literacy, xii, 9

B
Bell, Philip, 4, 5
Benchmarks for Science Literacy, xii, 9, 27, **29**
Berland, Leema, 4, 5
Bricker, Leah, 4, 5
BSCS, 30

5E Model, 19–20, 30, **31**
Bybee, Rodger, 4, 5, 19, 30

C
Carnegie Corporation, 7, 8
Checklist for gaining understanding of *NGSS*, 1–2
Code and title for standards, 11, **14**
Common Core State Standards (CCSS) in mathematics and English language arts, 1, 13, 19, 26
Connection box, 11, 13, **14,** 19, 23, 25
Course maps, 30
Crosscutting concepts, **xi,** 3, 7
    creating activities based on, 19
    discussion topics for examination of, 24
    5E instructional sequence with emphasis on, **31**
    in foundation box, 11, **14,** 19
    instructional integration of, 28–29, **29**
    resources on, 4–6
Curriculum
    making changes in, 30
    study group on the *NGSS* in relation to, 21–26

D
Development of the *NGSS,* xii, 3, 8–9
    NSTA's role in, 9
    partners for, 9
Disciplinary core idea(s), **xi,** 3, 7–8
    articulation across grade levels, 25–26
    connections to other disciplinary core ideas at this grade level, 25
    creating activities based on, 18–19
    discussion topics for examination of, 23–24
    5E instructional sequence with emphasis on, **31**
    in foundation box, 11, **14**
    instructional integration of, 27
    resources on, 4–5
Duschl, Richard A., 5

E
Engineering, technology, and applications of science, **xi,** 8, 13, **14,** 24–25, 32
    discussion topics for examination of, 24–25
    professional development for integration of, 32
English language arts, connections to *Common Core State Standards* in, 1, 13, 19, 26
Exploring the Science *Framework* and Preparing for the *NGSS* (journal series), 4

F
Financial resources for implementation of the *NGSS,* 33
5E Model, 19–20, 30, **31**
Foundation box, 11, 13, **14,** 23

G
Grotzer, Tina, 6
Gunckel, Kristin, 6

I
Implementation of the *NGSS*, 27–33
   changes needed for, 27–33
      in assessment, 32–33
      in curriculum, 30
      estimating amount of, 27
      in instructional materials, 30, **31**
      to integrate engineering, 32
      to integrate practices, disciplinary core ideas, and crosscutting concepts, 27–29, **28, 29**
      professional development for, 31–32, **32**
   course maps for, 30
   creating a plan for, 3
   developing support for, 33
   financial resources needed for, 33
   in limited trial situations, 3
   moving to instruction, 17–20, **20,** 21
   study group for planning of, 2, 21–26
   timeline for decisions about, 2
   workbook for, 27
Inquiry-based teaching, 28, **28**
Institute for Advanced Study Commission on Mathematics and Science Education, 8
Institute of Medicine, 9
Instruction, moving from the *NGSS* to, 17–20
   creating a description of desired results, 17–18
   design process for, 17, **20**
   developing activities from a standard, 18–20, 21
      assessment, 20
      crosscutting concepts, 19
      disciplinary core idea, 18–19
      integration of dimensions, 18, 19–20, 27–29, **28, 29**
      science and engineering practices, 19
   developing instructional materials, 30, **31**
   professional development for, 31–32, **32**
Introduction to the *NGSS*, 11–13, **12,** 11–15
   anatomy of a standard, 11–13, **12**
   definition of a standard, 13, 15

K
Kenyon, Lisa, 4
Koballa, Thomas R., Jr., 5
Krajcik, Joe, 4, 5

L
Learning goals, 3, 11, 15, 17–18, **20,** 27, 32
Learning progressions, 21, 25, 30
Lee, Tiffany, 4

M
Mathematics, connections to *Common Core State Standards* in, 1, 13, 19, 26
Mayes, Robert, 5
McNeill, Katherine, 5
Merritt, Joi, 4

Michaels, Sarah, 5
Milano, Mariel, 5
Miller, Zipporah, xii

N

National Academy of Engineering, 9
National Academy of Sciences, 8, 9
National Research Council (NRC), vii, xii, 3, 8, 9, 30
National Science Education Standards (NSES), vii, 27, **28**
National Science Teachers Association (NSTA), ix, xii, 1
   membership of, 9
   resources to support the *NGSS,* 1, 3–6, 8, 18, 33
   role in development of *NGSS,* 8, 9
   Science Anchors program of, 9
Nature of science, **xi,** 13, **14,** 22, 25
Next Generation Science Standards (NGSS)
   conceptual changes in, 1
   developing instructional materials based on, 30, **31**
   development of, xii, 3, 8–9
   getting started with understanding of, 1–3
   implementation of, 27–33
   introduction to, 11–15
   moving to instruction from, 17–20, **20**
   navigation of, ix–x
   NSTA resources for support of, 1, 3–6, 8, 18, 33
   organization of, 23
   path of progress in learning how to use, **xi,** xi–xii
   purposes of, 15, 27
   rubric to assess alignment of instructional materials with, 30
   study group on, 2, 21–26
   table of contents for, **xi**
   timeline for adoption and implementation of, 1, 2, 3
   use by state, regional, or district supervisors, xii
   use by teachers, xii
Next Generation Science Standards: Adoption and Implementation Workbook, 27
Next Generation Science Standards and the Life Sciences, 30
The NSTA Reader's Guide to A Framework for K–12 Science Education: Practices, Crosscutting Concepts, and Core Ideas, 4, 18

O

Ostlund, Karen L., ix

P

Passmore, Cynthia, 5
Performance expectations, 11, **14,** 21, 27, 32
   discussion topics for examination of, 24
Planning
   for implementation, 3
   for study of *NGSS,* 2–3, 21–26
Pratt, Harold, vii, xi–xii
Professional development, ix, xii, 3, 4, 20, 21, 26, 27, 30, 33
   designing a program for, 32, **32**

     to integrate engineering in curriculum, 32
     to modify instruction, 31
     standards and, 15
Project 2061, 9, 30

R
Reiser, Brian, 5
Riedinger, Kelly, 6
River, Ann, 5
Rubric to assess alignment of instructional materials with the *NGSS*, 30

S
Schwarz, Christina, 5
Schweingruber, Heidi, 5
Science, 9
Science Anchors, 9
Science and engineering practices, **xi,** 3, 7, 17, 18, 19, **20,** 24
     as assessed outcomes and instructional methodology, 28
     compared with Abilities of Inquiry from the *NSES,* **28**
     creating activities based on, 19
     discussion topics for examination of, 24
     5E instructional sequence with emphasis on, **31**
     in foundation box, 11, **14,** 19
     integration of, 27–28
     resources on, 4–5
Science for All Americans, 9
Science for the Next Generation: Preparing for the New Standards, 4
Shader, Bryan, 5
Smith, Deborah, 5
Sneider, Cary, 5
Standard(s)
     anatomy/content of, 11–13, **14**
     creating activities from, 18–20
     definition of, 13
     examining a page of, 23–26
          articulation of disciplinary core ideas across grade levels, 25–26
          connections to *Common Core State Standards* in English language arts, 26
          connections to *Common Core State Standards* in mathematics, 26
          connections to other disciplinary core ideas at this grade level, 25
          crosscutting concepts, 24
          disciplinary core ideas, 23
          discussion topics, 23
          engineering, technology, and applications of science, 24–25
          nature of science, 25
          performance expectations, 24
          report/summary of, 26
          science and engineering practices, 24
     example of, **12**
     organization of, 23
     professional development and, 15
     relation to instruction and curriculum, 13, 15
     study group for review of, 2, 21–26

supporting connection box for, 13, **14**
Study group, 2, 21–26
    to become familiar with the *Framework,* 22–23
    examination of a page of the standards by, 23–26
    membership and size of, 21–22, **22**

T

Taylor, Amy, 6
Timeline for adoption and implementation of the *NGSS,* 1, 2, 3
Title and code for standards, 11, **14**
Translating the NGSS for Classroom Instruction, 4, 20
Tsou, Carrie, 4

U

Understanding the *NGSS,* 1–6
    checklist for, 1–2
    collecting resources for, 2
    creating an implementation plan, 3
    determining state or district's plan for adoption and implementation decisions and timeline, 2
    following a study plan, 2–3
    forming a study group, 2, 21–26
    NSTA resources for, ix, xii, 3–6
    using in limited trial situations, 3
U.S. Education Delivery Institute, 27

V

Van Horne, Katie, 4, 5

W

Web seminars, 4, 5, 6, 33
WestEd, 30
Willard, Ted, xii
Workosky, Cindy, xii
Wysession, Michael E., 5